GT40 PHOTO ARCHIVE

Iconografix continuously seeks collections of archival photographs for reproduction in future books. We require a minimum of 120 photographs per subject. We prefer subjects narrow in focus, i.e., a specific model, railroad, racing venue, etc. Photographs must be of high-quality, suited to reproduction in an 8x10-inch format. We willingly pay for the use of photographs.

If you own or know of such a collection, please contact: The Publisher, Iconografix, PO Box 609, Osceola, Wisconsin 54020.

GT40 PHOTO ARCHIVE

Edited by Brian Winer and Wallace A. Wyss
Foreword by Bob Bondurant

Iconografix
Photo Archive Series

Iconografix
PO Box 609
Osceola, Wisconsin 54020 USA

Text Copyright © 1997

All rights reserved. No part of this work may be reproduced or used in any form by any means... graphic, electronic, or mechanical, including photocopying, recording, taping, or any other information storage and retrieval system... without written permission of the publisher.

We acknowledge that certain words, such as model names and designations, mentioned herein are the property of the trademark holder. We use them for purposes of identification only. Information contained herein is true to the best of our knowledge. This is not an official publication.

Books in the Iconografix *Photo Archive Series* are offered at a discount when sold in quantity for promotional use. Businesses or organizations seeking details should write to the Marketing Department, Iconografix, at the above address.

Library of Congress Card Number 96-78341

ISBN 1-882256-64-6

97 98 99 00 01 02 03 5 4 3 2 1

Printed in the United States of America

PREFACE

The histories of machines and mechanical gadgets are contained in the books, journals, correspondence, and personal papers stored in libraries and archives throughout the world. Written in tens of languages, covering thousands of subjects, the stories are recorded in millions of words.

Words are powerful. Yet, the impact of a single image, a photograph or an illustration, often relates more than dozens of pages of text. Fortunately, many of the libraries and archives that house the words also preserve the images.

In the *Photo Archive Series,* Iconografix reproduces photographs and illustrations selected from public and private collections. The images are chosen to tell a story—to capture the character of their subject. Reproduced as found, they are accompanied by the captions made available by the archive.

The Iconografix *Photo Archive Series* is dedicated to young and old alike, the enthusiast, the collector and anyone who, like us, is fascinated by "things" mechanical.

Bob Bondurant, racing great, posed with a Ford GT40 at Los Angeles International Airport, December 1964.

FOREWORD

The first two Ford GT40s were built in England at Ford Advanced Vehicles (FAV), with John Wyer in charge of the program. The cars ran three races in 1964, including the 24 Hours of Le Mans, where they ran fast but did not finish. Following the race, the cars were shipped to Carroll Shelby in the United States. Shelby transformed the cars: reworking the suspension; adding larger brakes; putting a more aerodynamic front end on the car; and changing to a Cobra iron block V-8, from the original alloy block Indy engine. Shelby kept the Italian Colotti gearbox, but this was later replaced with a German ZF gearbox. The man in charge of the work was Phil Remington. Phil was fantastic. I learned a lot from Phil. He believed that there was nothing that could not be done—that there was always a way to solve a problem.

We at Shelby took the cars, evaluated them, and spent many hours testing them at Riverside Raceway and Willow Springs. We got the cars very competitive and then raced them at Daytona in 1965. I qualified second behind Pedro Rodriguez, who beat me in a Ferrari. During the race, we ran very, very well. We had problems with the electrics during a pit stop and dropped back, but Ritchie Ginther and I charged ahead and finished third. Ken Miles and Lloyd Ruby won the race in their GT40 MK II. It was after this race that we switched over to ZF gearboxes.

The next race in which I drove a GT40 was the Targa Florio in Sicily. Sir John Whitmore and I shared an open version, fitted with all the Shelby updates, and entered under the banner of FAV. We led for a while, but had a problem when John lost a front wheel. He eased the car off the road without damaging it, but a spectator stole the knockoff as a souvenir. Our pit crew sent out a runner with a replacement, only to find that the spectator had sheepishly returned with the original. We remounted the wheel, and I took over the car. We had dropped way back. I was trying to make up the time lost, when I rounded a corner at about the 12-kilometer marker and hit gravel. You have to understand, when the Sicilians repaired a road, which was done about three days before the race, they merely tamped asphalt down to fill the holes. After four or five laps of the race, the asphalt came up. The corner was a blind right-hander. I hit the loose asphalt, started to lose it, slid off the road and headed for the kilometer marker, which stuck up about two-and-one-half feet from the surface of the road. I hit that marker so hard that the car lifted off the ground and flipped sideways. The car slid on its right side and, as it was a right-hand drive car, I was sliding along looking at the road just inches from my face. The impact was so hard it broke my seat belt. I was just hoping and praying that the car would not roll all the way over. Luckily it did not. It slid to a stop. A group of very enthusiastic Sicilian spectators gamely volunteered to right the car, but the damage was too great. Instead, I walked back to the last town through which I had passed, went to the local bar, ordered a beer and waited for the race to finish.

My next race driving the GT40 was at Le Mans in 1965, where I shared the driving with Umberto Maglioli. My small block 5.3-liter, Shelby-prepared car was entered under the banner of Rob Walker. We turned third fastest in qualifying. Phil Hill was fastest in a 7-liter GT40; Surtees second in a Ferrari 330P2; the late Bruce McLaren fourth in a 7-liter GT40 MK II roadster, after which it was Ferraris and GT40s on down the line. In qualifying, I turned 212 mph down the Mulsanne Straight. The car felt fantastic, very stable. Shelby, who had worried that the 5.3-liter engine was too fragile, swapped it for a 4.7-liter iron block. Unfortunately, the car was out of the race in about two-and-one-half hours due to overheating. Ford did, of course, get the GT40 going after that, and came back with an onslaught in 1966 and won Le Mans. The GT40s also won at Le Mans in 1968 and 1969.

The GT40 was a very advanced automobile. It was one of the best race cars ever built. In 1966, I drove a works car for Ferrari. Although the Ferrari P3 was very good, the GT40 was much easier to drive. It was just that much quicker, which allowed it to win. By beating Ferrari, in both the GT and prototype categories, Shelby and Ford advanced auto racing. When I think of my association with the GT40s—of all that test driving, of running at Le Mans at 212 mph, and, yes, even the Targa Florio incident—I feel immensely proud of the Ford Motor Company and Shelby America and the accomplishments of this great car.

I hope you enjoy this book.

Bob Bondurant, President and Owner
The Bob Bondurant School of High Performance Driving
Firebird International Raceway
PO Box 51980, Phoenix, Arizona 85076-1980 USA
(800) 842-7223; (520) 796-1111; http://www.bondurant.com

One of the first three Ford GT40s, as photographed in 1964. Many design changes followed.

INTRODUCTION

The origins of the Ford GT40, unlike those of the AC Cobra, cannot be claimed by two countries. Although the GT40 was built in England, as was the Cobra, the impetus behind the car came from Dearborn, Michigan USA—and from the ambitions of Henry Ford II.

In 1963, Mr. Ford sent emissaries to Italy with open checkbooks to buy Ferrari. After a few weeks of negotiation, Enzo Ferrari abruptly canceled the talks, perhaps fearing that, if the deal went through, the Ferrari name would be diminished. After getting the cold shoulder from Ferrari, Ford issued new marching orders to his troops—Ford Motor Company would make its own sporty cars and beat Ferrari in the most prestigious race of them all, the 24 Hours of Le Mans.

In Dearborn, Ford's designers tackled the assignment in their usual way—making drawings, then sculpting clay models. But this was clearly not going to work. In Detroit, it could take three years to move from drawings to a finished car, too slow a pace for racing. Ford's Roy Lunn, who engineered the mid-engine Mustang I show car, approached Cooper and Lotus, neither of which at the time could act as Ford's subcontractor on a new mid-engine car. As fate would have it, Eric Broadley, founder of the English firm Lola Cars Ltd., built a small mid-engined car in 1963, using a 4.2 liter Ford V-8 borrowed from Carroll Shelby. He entered his car at Le Mans. Although he crashed, Ford recognized that Broadley's car met many of their criteria for a winning, mid-engined car. Ford promptly hired Broadley, bought two of the three Lola GTs he had built for use as chassis development mules, and formed a new company, Ford Advanced Vehicles, to build Ford GTs in England. By April 1964, they had two cars ready to test at Le Mans. It proved a brutal sorting out. The cars were ungodly fast but dicey at speed. It quickly became apparent that there was a lot to learn when running at speeds approaching 200 mph. That first racing season the team fell flat on its face in three successive events, even with drivers the caliber of Phil Hill, Bruce McLaren, and Ritchie Ginther.

In his foreword, my friend Bob Bondurant tells how Ford turned to Carroll Shelby, a bona fide Le Mans winner, for help. (Bob does not get into it, in the short space we gave him, but Ford hired other experienced racing organizations to help them, as well—including Holman & Moody of Charlotte, North Carolina, the stock car racers, and Alan Mann in England, who had worked with Ford when they sent the Falcon Sprints to Europe in the pre-Mustang days.) Shelby was invited into the Ford GT program just as Ford had decided to follow Shelby's Cobra evolution and stuff the 7-liter V-8 into the GT. Shelby's reasons for going to the 7-liter in the Cobra had been mostly marketing, his response to Corvette's 396-cubic inch V-8. However, Ford's reason for going to the big block in the GT was reliability. The big block Ford had years of NASCAR racing behind it and was bulletproof. It did not hurt that it produced almost 100 horsepower more than the small block 289.

The "big block" Ford GT40 MK II was not an immediate success. Ford's effort again fell flat at Le Mans in 1965. This time, it was a problem with transmissions. A very succinct order came down from the top floor at Ford World Headquarters, "Win or else!" In 1966, it all came together. At Le Mans, Ford followed the philosophy "safety in numbers". There were more than a dozen Ford GTs on the grid, and the end results were everything Henry Ford II had hoped for—a 1-2-3 finish. Ford's image throughout the world was suddenly cranked up a few notches.

Just to show that the 1966 victory was no fluke, Ford bankrolled another year of endurance racing. With an all new car, Ford won again in 1967. Was that the end? Not yet. John Wyer, who had managed Ford's Le Mans effort, went off on his own and, on a shoestring budget provided by his sponsor Gulf, won Le Mans in 1968 and 1969—with the same damn car! Wyer's success marked the vindication of Ford's original concept, proving that a big block was not needed to win Le Mans. Moreover, he proved that the original MK I, even when classed as a production sports car, could beat the prototypes.

There will be those who would argue against our having included the MK IV in this book, because the MK IV is not descended from the MK I and MK II. But it was developed by the same engineers, run by the same teams, piloted by many of the same drivers, and was part and parcel of Ford's efforts at Le Mans.

I am gratified to have this opportunity to share my collection of GT40 photographs. I confess I was not paying attention when Ford was winning those races. Blame college, my burgeoning career in advertising, and the US Army. I became a GT40 fan only after authoring *Shelby's Wildlife: the Cobras and the Mustangs*. After completing that book, I realized that Shelby's involvement with Ford did not end with the Cobra. There was more story to tell. Here is the Ford GT40 component of that story in pictures.

Wallace A. Wyss
May 1996

I attended college in Florida during the mid to late 1960s, and was thus able to see several Daytona 24-Hour and Sebring 12-Hour races first hand. I must admit, when I attended those races, the GT40s did not attract my attention. A decade later, when I began to collect Shelby memorabilia, I came to the realization that the Shelby story included that of the GT40.

When we met, Wallace Wyss and I shared an interest in mid-engined cars, and soon we both began collecting material on the Ford GTs. I finally went one better and put in an order for a Safir Engineering continuation car, built with the blessing of JW Engineering, builder of the original GT40. When Wallace mentioned that Iconografix was interested in presenting our collections of photographs to the public, we set to work to organize our files chronologically, so that we could tell the story of the Ford GT in pictures from start to finish. We consider this book to be a tribute to the Americans and Britons who conceived these cars, and a tribute to the men who raced them. We are both particularly grateful to Ford Motor Company for having put the car in the showrooms, where the public could buy it. Even today, you occasionally see a Ford GT40 on the street. To see one—to hear one—is an experience, even 30-odd years after Ford conceived it. That is why this book exists.

Brian Winer
May 1996

A clay model of October 1963. The nose was remarkably simple. Location of headlights was not yet fixed. Although barely visible here, the triangle-shaped hot air exhaust vents were already in place.

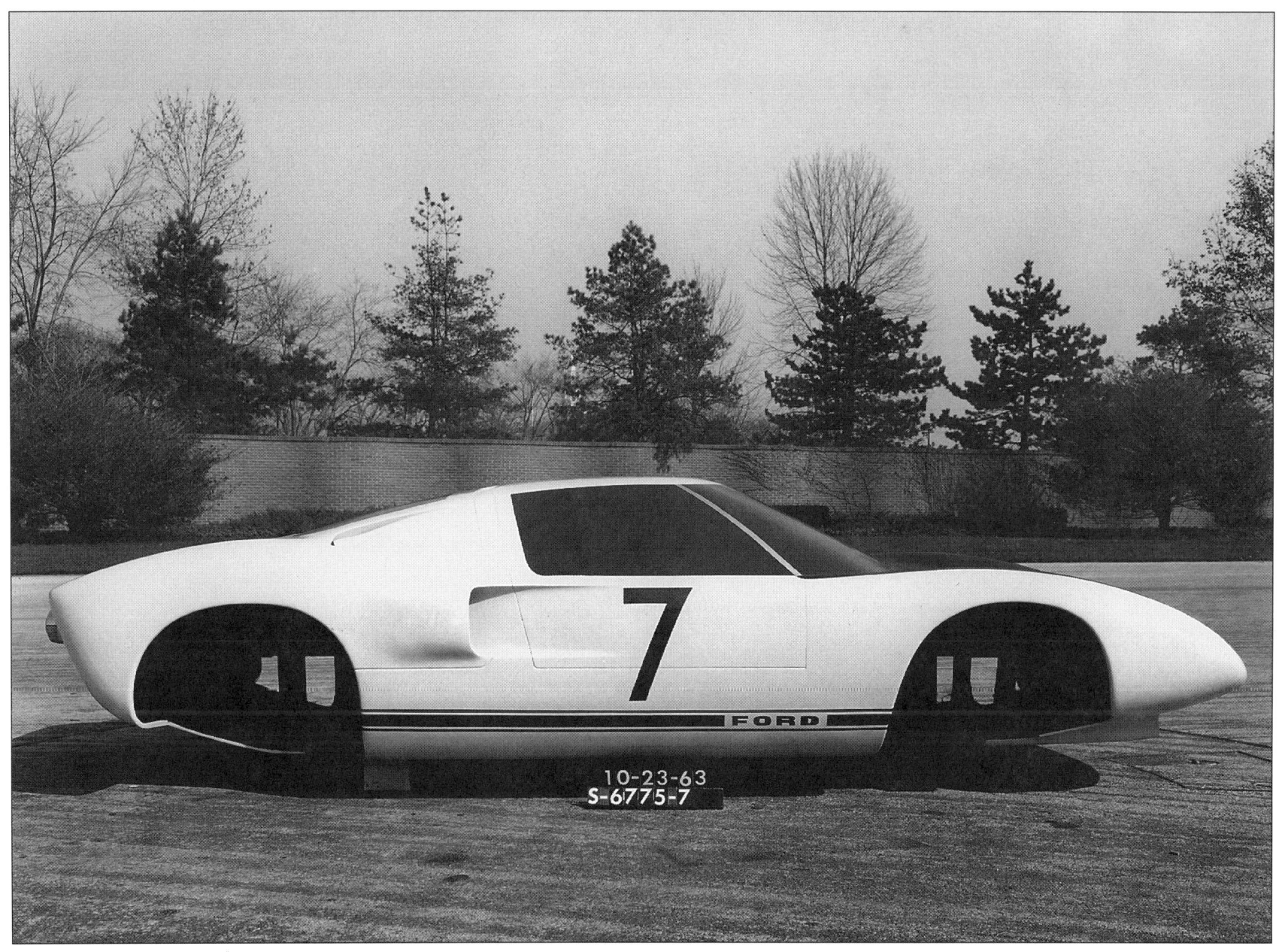

In this photograph, the A-pillars look deceptively thin. Perhaps they were covered by the tinted side window. The final car would strongly resemble the clay model.

This view shows the ingenious rear upper air scoops. The louvers at the rear were eventually dropped in favor of screened vents, which could exhaust more hot air. The elaborate cover over the transaxle and the chrome-plated exhaust pipes may have suited a road car, but neither were suited to the track.

An unidentified Ford official stood by the mockup, showing just how low the 40-inch car was. The finished car weighed 1825 lbs. without fuel.

The Ford GT of 1964, a giant step ahead of the Lola GT from which it evolved. As complete as this car might have appeared, many things would change in 1965. Among the changes: wire wheels would go, because they lacked rigidity, were heavier than mag wheels, and were not available in the desired rim width; larger diameter exhaust pipes were used; rear deck ventilation slots were added; and, most importantly, a sharp-edged spoiler "lip" was molded into the rear deck. Ford had found that the rear end of the car lifted at speed, during April 1964 tests at Le Mans. Both test cars crashed but the drivers were unhurt.

This photograph of a lone prototype being unloaded from a single-car trailer illustrates how small the Ford GT project was when it began. Eventually, more than $20 million was spent toward the goal of winning Le Mans. This may even be the first car built.

Phil Hill, hand on door, contemplated the task ahead—the Ford GT's first race, the Nürburgring 1000 Km, May 31, 1964. A pensive Eric Broadley, at left, creator of the Ford-powered Lola GT, looked on. One of the first three prototypes, this car completed only 15 laps before a weld in the rear suspension failed. Note how the need for more air to the radiator necessitated cutting an air intake into the front deck. Originally, a pair of road lamps were mounted in the spoiler. They were removed, mounted below the headlamps, and enclosed behind Plexiglas.

Although the GT40 body was constructed from fiberglass, ductwork up to the hood vents was aluminum. The front and rear body sections could be removed in one minute.

In 1964, the GT40 impressed the crowds at Nürburgring with its speed. The first year, Ford ran their pushrod, dry sump, alloy block Indy V-8. It displaced 255.3-cubic inches and produced 350 hp at 7200 rpm with 294 lb ft torque at 5600 rpm.

A 1964 prototype at the track. The car performed well beyond the Cobra Daytona coupe's ceiling of 170 mph.

This view shows how narrow the tires were in 1964. Borrani wire wheels were still being used at this stage, Shelby not yet having selected a supplier of solid wheels. Ford found out what Ferrari already knew—the high side loads generated by a mid-engined car could snap spokes on a wire wheel. Also note the carburetors, four 48mm twin-choke Webers.

Le Mans 1964. The first five cars off the line were Ferraris. A Ford GT, co-driven by Masten Gregory and Ritchie Ginther, sat on the starting grid. Car No. 3 (behind Aston No. 18) was an AC Cobra coupe that crashed during the race, killing three spectators. Phil Hill and Bruce McLaren set the lap record but withdrew with gearbox trouble. After clocking 191.4 mph on the Mulsanne Straight, Ginther also withdrew with gearbox trouble.

Le Mans 1964. The Schlesser/Attwood entry hounded by a Zagato-bodied Maserati 450S. The GT40 was running sixth, when it caught fire and burned on the Mulsanne Straight.

The chassis of the Len Bailey-designed Ford GT was a strong unitized tub. Shown is a fire-damaged MK I being restored in the 1980s by John Collins, a former Shelby-American International (SAI) employee. Designers made no attempt to pass along any loads to the front or rear of the body. Carroll Smith, the engineer/race car consultant who worked for Shelby, states that the GT40 frame was "one of the strongest race car chassis built, up to that time, with over 20,000 lb ft of torsional rigidity." Nevertheless, when Ford tried to build targa models, lack of rigidity was a problem. It did not deter the targas from winning, however, as one did at Sebring in 1966.

MK I Chassis No. 104 at Daytona, February 1965. Drivers Ginther and Bondurant finished third overall. In addition to the new nose and tail, this car showed a few subtle changes from the original prototype: a front spoiler was added; a half-scoop to the rear roof section resulted in more air being funneled in; rear fenders were wider; Halibrand mags in place of wire wheels. Sponsors included Goodyear and Autolite. The Daytona Cobra coupe-style twin white stripes indicated a Shelby-American car.

Another shot of the Daytona MK I. Front jacking "handles" projecting from the nose permitted use of "quick-lift" jack. Bondurant, the taller driver here, talked with co-driver Ritchie Ginther.

A MK I roadster. Wire wheels usually indicated an entry from Europe, as Shelby's crew disdained them. The first GT40 roadsters used a nose that looked more like the 1964 car, rather than the "full" definitive MK I nose. According to Carroll Smith, "an enormous sacrifice in rigidity resulted when we made some coupes into roadsters." Ironically, after the GT program was concluded, Ford gave one of the four MK I roadsters to a movie stunt man and cut another up and threw it into a dumpster!

A 7-liter GT40 roadster at the Arizona test track, probably Spring 1965. This may have been the car that became the X-1, used by McLaren for Group 7 racing and later by Ford as a MK II sports prototype. If so, it won at Sebring in 1966. Note the see-through Plexiglas rear spoiler. Frequently Kar Kraft, Ford's in-house race car building subsidiary, experimented on their own and sent the results to Shelby, who had power to accept or reject their changes. According to Carroll Smith, the derisive name for Kar Kraft among Shelby employees was "Kiddy Kar".

Probably the car used by the press car fleet in England, a GT40P—"P" for "production". Note Borrani wire wheels and Weber carburetors (velocity stacks barely visible). In America, a single four-barrel was optional. Equipped with Webers, the GT40P would do 0-60 mph in 5.3 seconds. The road cars had softer brake linings and 25% softer shocks and springs.

Interior of the GT40P road car. Note road car details like door map pockets and leather upholstery. Base price was $17,000; up to $20,000 with extras.

This was the definitive shape of the MK I. Although a road car, this unit had more than one taillamp per side, as was preferred by racers. Since rules required that racer cars had at least one operable taillamp, the more fitted to your car the lower your risk of being "black-flagged". The fact that this car had Michigan license plates meant that it was probably driven home at night by executives. This was probably—after sampling the 427 Cobra press cars—the biggest thrill of their careers at Ford.

The road car's luggage area—two heat-resistant aluminum boxes closed with Dzuz fasteners. The 289 engines were rated 335 hp at 6250 rpm. Note special silencers for the road version. This photo dated December 12, 1965.

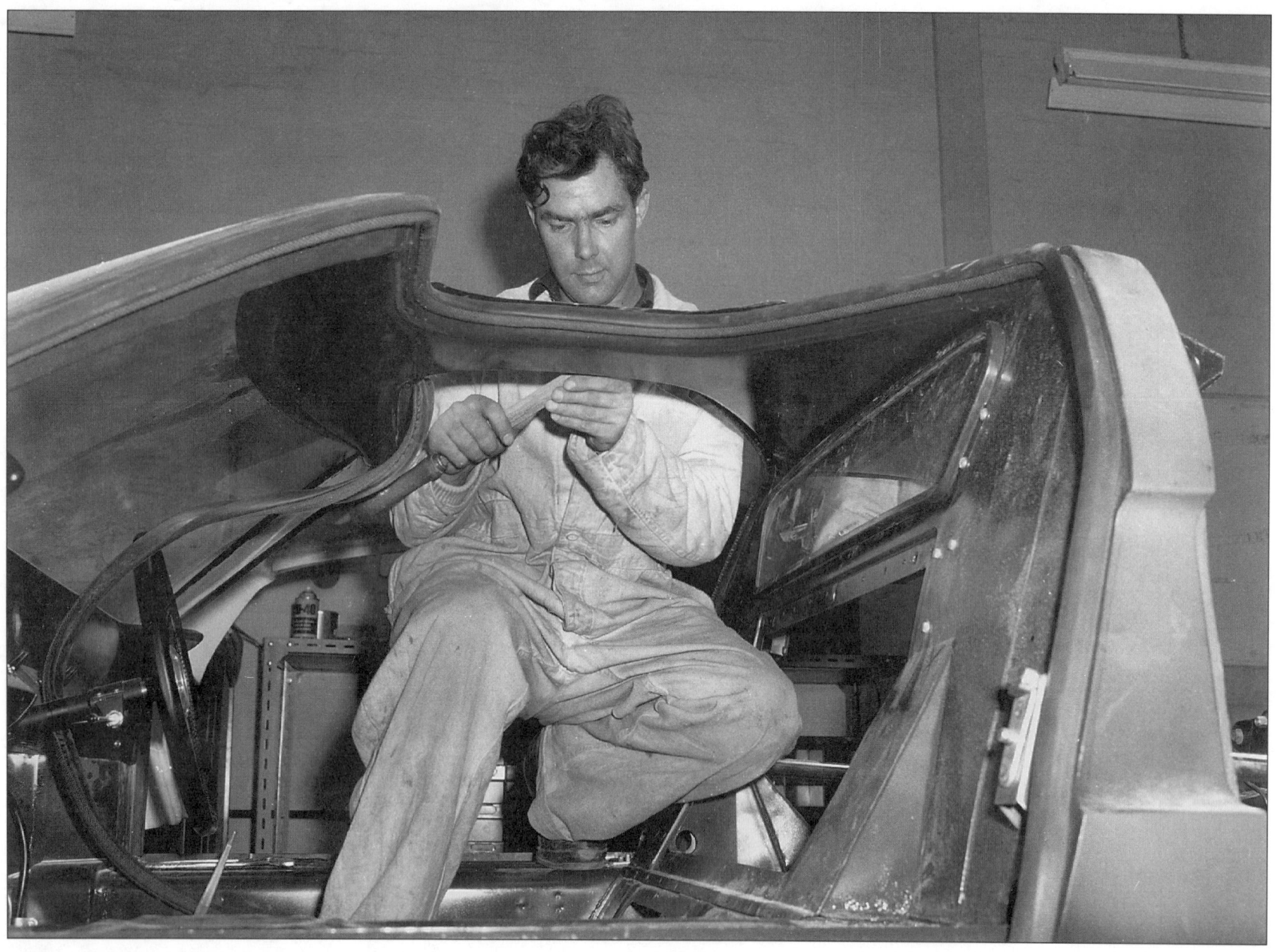

Constructing the GT40. Although some large stampings were used, much of the car was assembled of small bits. Here a worker installs weather-stripping on the roof.

Ford Advanced Vehicles (FAV), Slough, England, where GT40s were "mass produced". To be classed by F.I.A. as production sports cars, the class in which the GT40 was most competitive, it was required that Ford build a minimum of 50 cars. Here, GT40Ps await pick-up by a transporter. Some were shipped to the USA, where they miraculously slipped in under the wire of new DOT safety regulations.

Front view of the Noel Edmonds car, which was finished in lime green metallic with a dark green stripe. Glass windows popped open for ventilation. Note that quick-lift jacking hooks were absent.

A GT40 "kitted out" for the road. This car was sold to Noel Edmonds, a British disc jockey. He added chrome window frames from the MK III and full leather interior, although using the same ringlets from the cloth interior (see page 40). Wheels were BRM 6-spokers from the later Gulf GT40s.

Engine compartment of the Edmonds GT40. Many road car owners ran a four-barrel for convenience but this owner wanted the Webers. Note luggage box and metal shields over the exhaust to protect the fiberglass.

The dashboard of the Edmonds car. A 200-mph Smiths speedometer was added. Up-dates included a fire extinguishing system and air conditioning.

The start of Sebring 1965. It would be generous to say that the safety rules were a bit relaxed in the Sixties. A few scattered hay bales were all that separated spectators from a swarm of 200-mph cars jousting for the lead. A Chaparral roadster appeared to be dusting off Ken Miles in GT40 No. 11. A Ferrari 250LM driver lost precious time trying to shut his door properly. Miles and McLaren finished second in their 289-equipped MK I.

Miles (checked hat) talks to McLaren in the pits at Sebring. Note football player-style helmet popular before full-coverage helmets were introduced.

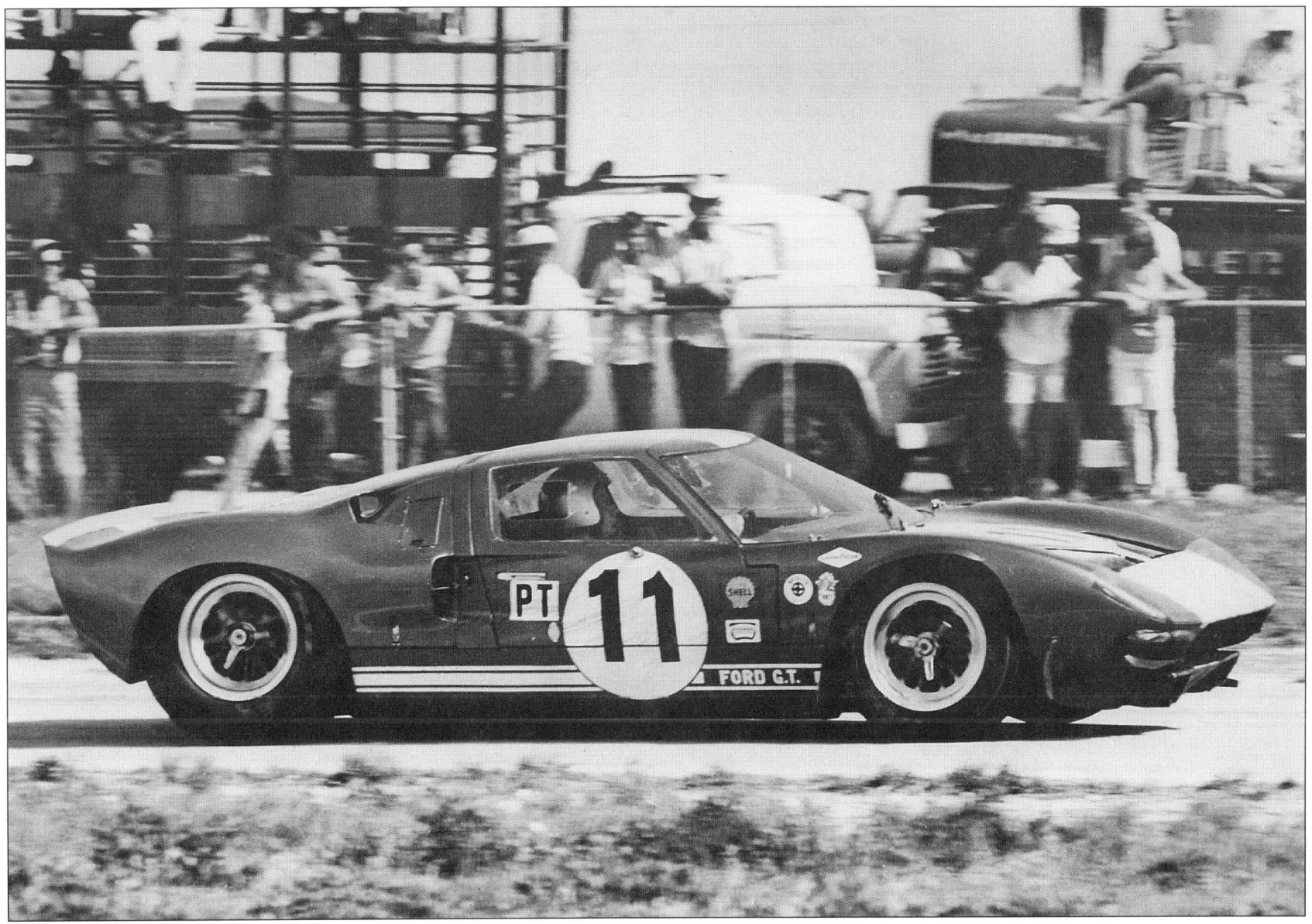

McLaren at the wheel of Chassis No. 104. Note the crude front spoiler with air intakes at each end, and rear spoiler, to which it appears a sharper lip was added. Hogged-out rear roof air intake and crude access door are visible. Sponsor decals applied to the dark blue body included Goodyear, Shell, Autolite, and Rebestos.

The 1965 Targa Florio. Bondurant behind the wheel of the FAV-sponsored MK I roadster. As recounted in his foreword, Bondurant crashed the car on the ninth lap.

For the 1965 season, the small block engine—despite its success in the Cobra—was judged too unreliable. They took measurements and found that their old NASCAR favorite, the 427 big block, would fit. It weighed 260 lbs. more than the smaller block, and dictated that the car's driveshafts, suspension, and chassis be beefed-up.

When Ford changed to the big block, they tried to improve the car's road-hugging characteristics by lengthening the nose. Here Ford designers dickered over surface details, as the wind blew the wool tufts around in a "tuft test" of the 1965 "long-nose" big block MK II.

A new MK II parked outside of Kar Kraft in Allen Park, Michigan. Major changes from the MK I included: addition of front fender side air vents; longer nose; single hood vent, in place of the pair of triangular-shaped vents.

For some reason, there are few pictures of the 1965 MK II. Perhaps, because it failed in its mission to win at Le Mans. Concerns about directional stability were addressed through the addition of side nose canards and foot-tall tailfins. The vent behind the front wheels allowed trapped air to escape the wheel wells. Carroll Smith states that the nose of the MK II was too long and that it "packed in air underneath." This was the Ken Miles/Bruce McLaren car at Le Mans. The car was white, with metallic gold-painted mags, black hood and black side stripes.

Detail of the Marchal headlamp on a 1966 MK II, same for a 1965 MK I. Note how deep the cavity of the front hood exhaust vents were by 1965. The idea of covering the hood vents with a grate was dropped after the 1964 season.

Le Mans 1965. "Where did you say you want these cars delivered, Mister?" What a difference from the 1964 season when a one-car trailer was all that was required. This rig carried four coupes, two small-block MK Is on the top deck and two big-block MK IIs on the bottom, one roadster, and a Cobra Daytona coupe. Among the group, the Cobra was the sole car to finish the race.

Race number 10, Chassis No. 104, at speed at Le Mans 1965. Ford was plagued with overheating problems in the small blocks; transaxle breakage in the big blocks.

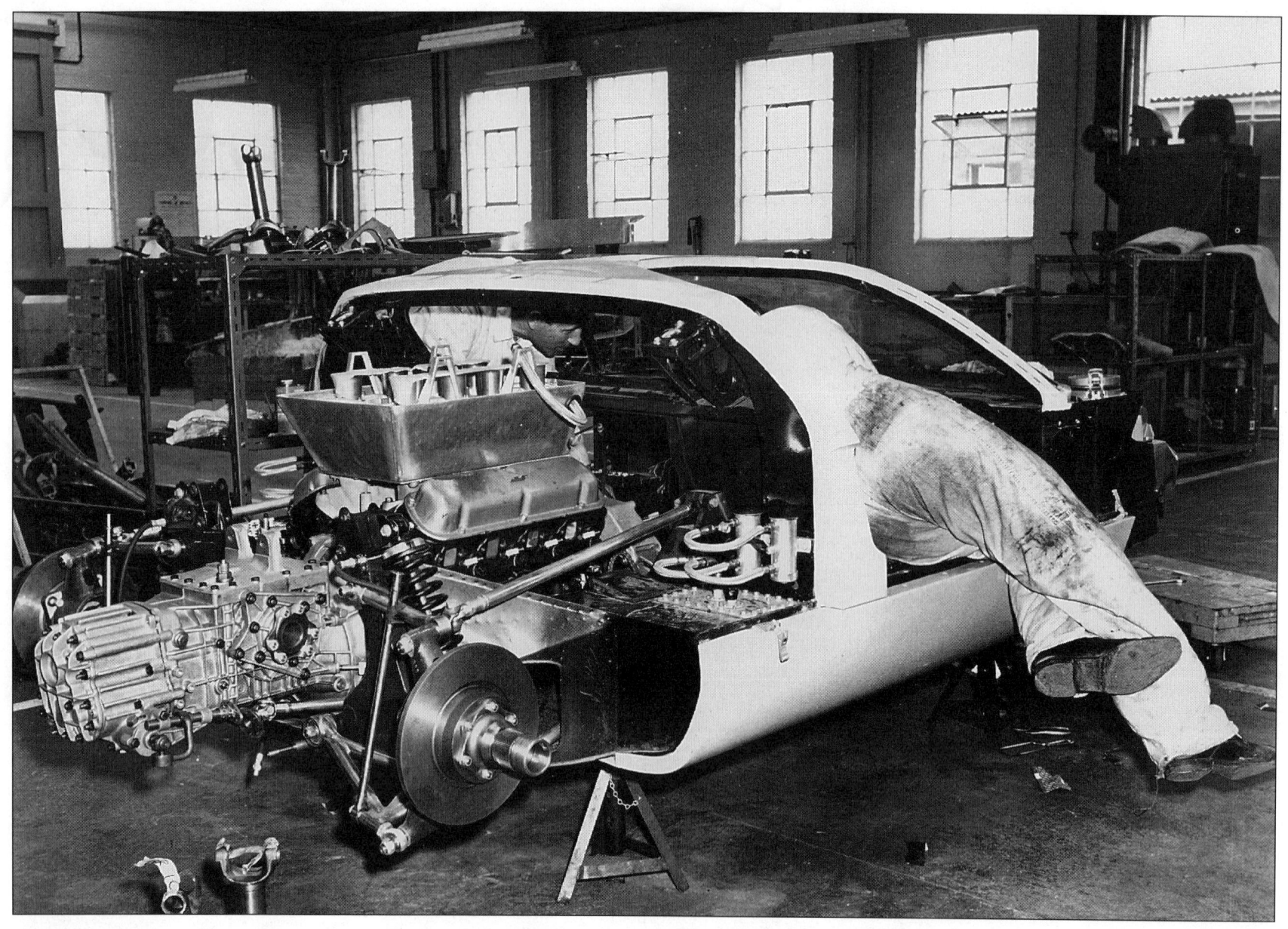

A small-block in the FAV shop at Slough, England. The Webers were enclosed in an aluminum air box, isolated from the heat of the engine. This car has had the tray, which stopped foreign objects from dropping into the carburetors, removed. Why solid discs? Ventilated discs were not yet available.

Bruce's Blunder? Hardly! New Zealander Bruce McLaren talked Ford into bankrolling this project, Chassis No. 110. Abbey Panels, the subcontractor who fabricated the GT's chassis, used their standard tooling to make a tub out of aluminum, instead of steel. McLaren chose to run a Hewland LG500 four-speed transmission, instead of ZF, because it permitted a wider choice of axle ratios. He also grafted on a 1965 Le Mans-style long nose. McLaren entered the car in a series of Group 7 races with Chris Amon as his driver. According to Carroll Smith, the money McLaren made on the deal helped bankroll his own race car company.

From the rear, McLaren's GT X-1 gave few clues to its GT40 heritage. He ran the car at Mosport, Riverside, and Nassau. Amon's best finish was fifth in the Times GP at Riverside. Even with its aluminum tub, the car was still too heavy to be competitive.

The interior of the X-1 was pretty neat, considering the pieced-together look of the rest of the car. The ringlet-style cloth upholstery is still there, as well as the original style steering wheel with its red, white, and blue center badge.

The X-1 showed efforts to eliminate the "divider" that had split the bonnet vents. The change reflected the notion that anything that got in the way of air, as it exited the radiator, was an impediment.

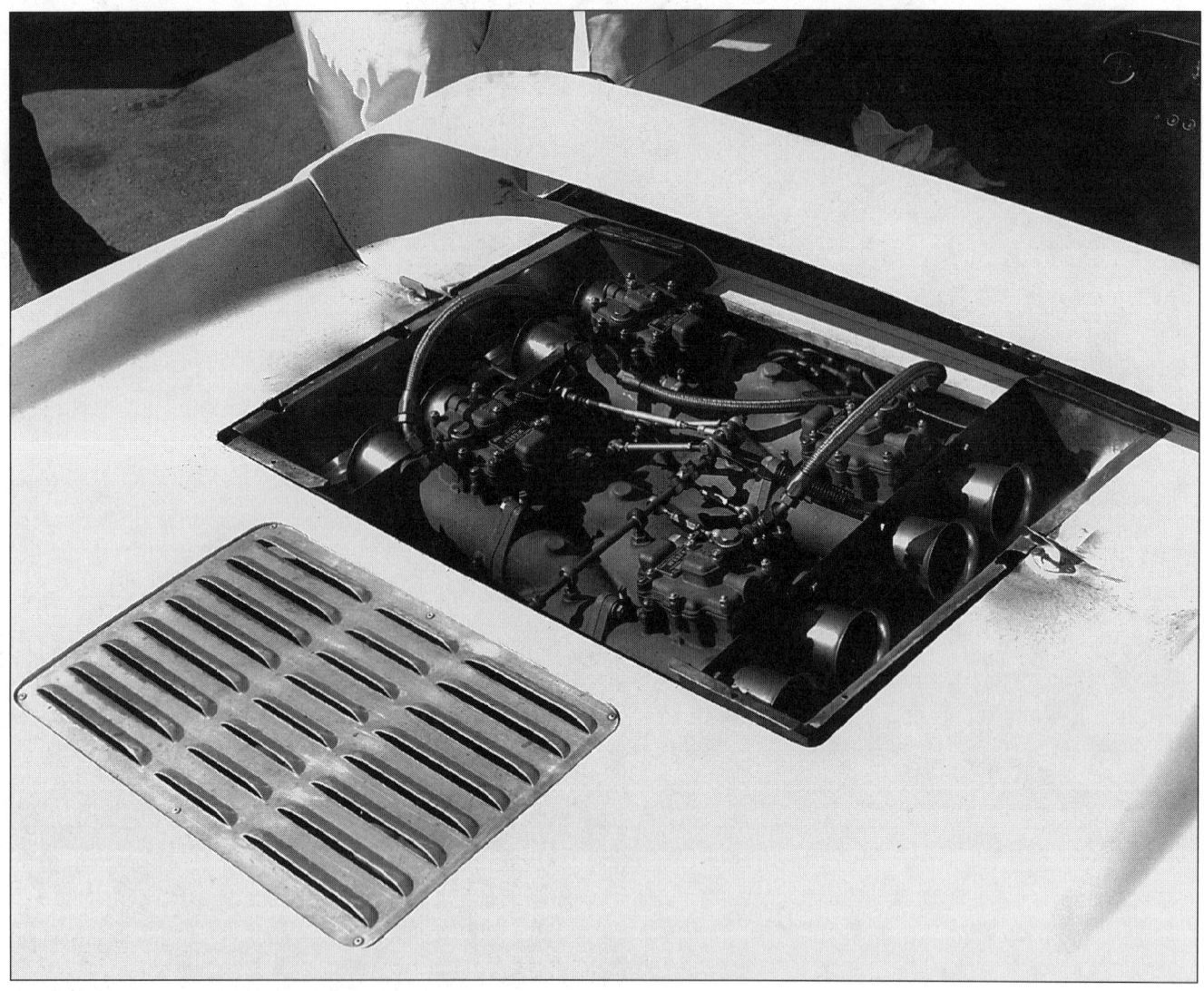

The engine of the X-1 was a thundering big block, rated at 485 hp with a single four-barrel. It produced something like 590 hp, with an Iso Grifo-A3C-style cross-ram manifold toting Webers. Carroll Smith states, "We could have gotten more power out of the big block, if we would have been allowed to use [Webers] at Le Mans. For some reason, Ford insisted on using American-made Holleys." Can-Am racers did not stay long with carburetors, turning instead to fuel injection.

During the second season of the big block, Shelby again made styling changes. The MK II was fitted with the definitive MK I style "short nose". A second set of scoops was added above the existing side scoops. Two more "snorkel" scoops were added to the rear deck. The slot in the body side took in air to cool the fuel pump. This model with four side scoops was referred to as the MK II-A.

April 1966. A MK II-A rolled out onto the Romeo, Michigan test track. It topped 201.5 mph and hit 210 on the straightaway. Neither front spoiler nor nose canards were thought necessary, although an aluminum spoiler was fitted to the rear. Nose was deeper to accommodate fog lamp plus turn signals in sealed pod. Hood vents had lost their triangular vents and were deeper, with a thin divider left for rigidity.

Metal truss at the rear did not hold up the gearbox but provided a means of mounting the rear body jacking hooks for the quick-lift jack. Note air vent on front fender. MK II-As of 1966 had quick-change ventilated discs.

MK II-A detail: (upper left) hood vents were still divided; (upper center) gas cap was flat-topped, square-edged, pop-open type; (upper right) locking pin that helped secure rear body section in position; (lower left) oil cooler, one of two on either side of engine. Fuel lines were aviation-type, braided stainless steel; (lower center and right) single four-barrel and twin four-barrel carburetors were tried on the MK II-A. This was the single set up, on what was probably the first MK II-A prototype. Engine valve covers were labeled "Experimental".

Daytona 24-Hour 1966. The race started late afternoon, but it looks as if this photo was shot in the early morning. This was Hansgen/Donohue, who finished third overall. Their car had a chassis (No. 1031) supplied by FAV; engine and body installed by Shelby-American, with final race preparation completed by a team of mechanics from Shelby and Holman & Moody. Note how part of the grille cavity was blocked off, an indication that the team manager thought the car was running too cool during the night. Non-reflective matte black hood was used on first prototype Ford GTs and brought back early in 1966 by some teams.

The Skip Scott/Dr. Dick Thompson MK I passes Ford's nemesis, a Ferrari, at Daytona 1966. Both cars were independents. The Ford, Chassis No. 1026, was run by Essex Wire. It did not finish.

1966 Daytona winners, Lloyd Ruby and Ken Miles (arm around the beauty queen). Carroll Smith, wearing a cowboy hat, was visible behind Ruby. Daytona was a clean sweep for Ford. MK II-Bs finished 1-2-3.

For 1966, Ford ran a single Holley double-pumper carburetor. Clear plastic was used to permit an unobstructed view rearward. Engines arrived ready-to-go from Ford's Engine & Foundry Division (their decal can be seen on the valve cover). Shelby's crew dyno tested every engine, if time permitted.

The chassis of the MK II was identical to that of the MK I, in many ways. With body panels removed, it was difficult to tell them apart. We believe that these were MK II-As being readied for Sebring 1965. Note how the radiator was tilted forward. Aluminum duct work was designed to carry the brake cooling hose from grille cavity to brakes.

Sebring, March 26, 1966. Prior to the race, Dan Gurney's MK II-A (foreground) received some final attention. The roof blister or "Gurney Bump" was fitted to all cars driven by Gurney. Chassis No. 110, No. 1 for the Sebring race, appeared shorn of its front and rear body sections. It would win the race.

Gurney, in white driving suit, appeared anxious, as he waited for his MK II to be readied in the pits. To his left, Carroll Shelby, in black shirt, stood at the rear of the car.

In a great come back for an old used car, the ex-Bruce McLaren failed GT X-1 was brought back into team car status by Shelby-American. Fitted with a MK II rear body section, a new short nose, and a tall glass windscreen, it was sent off to Sebring for the 12-Hour. Although it has been stated that this car was equipped with auto gearbox, according to team manager Carroll Smith, that never happened. The car won Sebring "by surprise", after Gurney's MK II coupe, which had led the race for 11 hours, broke down a half mile from the finish.

Sebring 1966. The advantage of a roadster configuration is that, when you win, everyone can sit on the targa roof bar while they accept congratulations. In this photo, Ken Miles was interviewed as the race queen looked on. Shelby, in tri-cornered black cowboy hat, sat between the queen and Miles co-driver, Lloyd Ruby.

How to win at Le Mans? Enter as many cars as you can. In 1966, there were 13 Fords vs. 14 Ferraris. We count eight GT40s in this photo, the majority were MK II-As. This was well before the race started, as seen by the emptiness of the stands. Car No. 1, in front, was the Miles/Hulme car that finished second overall.

Another view of the MK II lineup. Ford dumped the black hood/white body scheme and picked up the colors from the Mustang palette. No. 8 was yellow; No. 1 silver; No. 6 dark blue. Gold, maroon, red, black, and light blue followed down the line.

The Miles/Hulme MK II, a light sky blue. The quick-lift jack hooks, poking out of the grille cavity, were red and matched the striping. Center stripes were white; wheels gold; knockoffs were green on the right side of the car and red on the left, preventing any mix-up. Mechanic John Collins entered the car. Collins later restored GT40s.

This was the Gurney/Grant MK II, Chassis No. 1047, at Le Mans 1966. The car retired after 17 hours. Note Bruce McLaren seated on the wall. Denis Hulme stood at the front of the car. The man in the suit sporting sunglasses was Ford racing boss Jacques Passino. The man in the Stetson was Shelby team manager Carroll Smith, who went on to become a successful race team consultant.

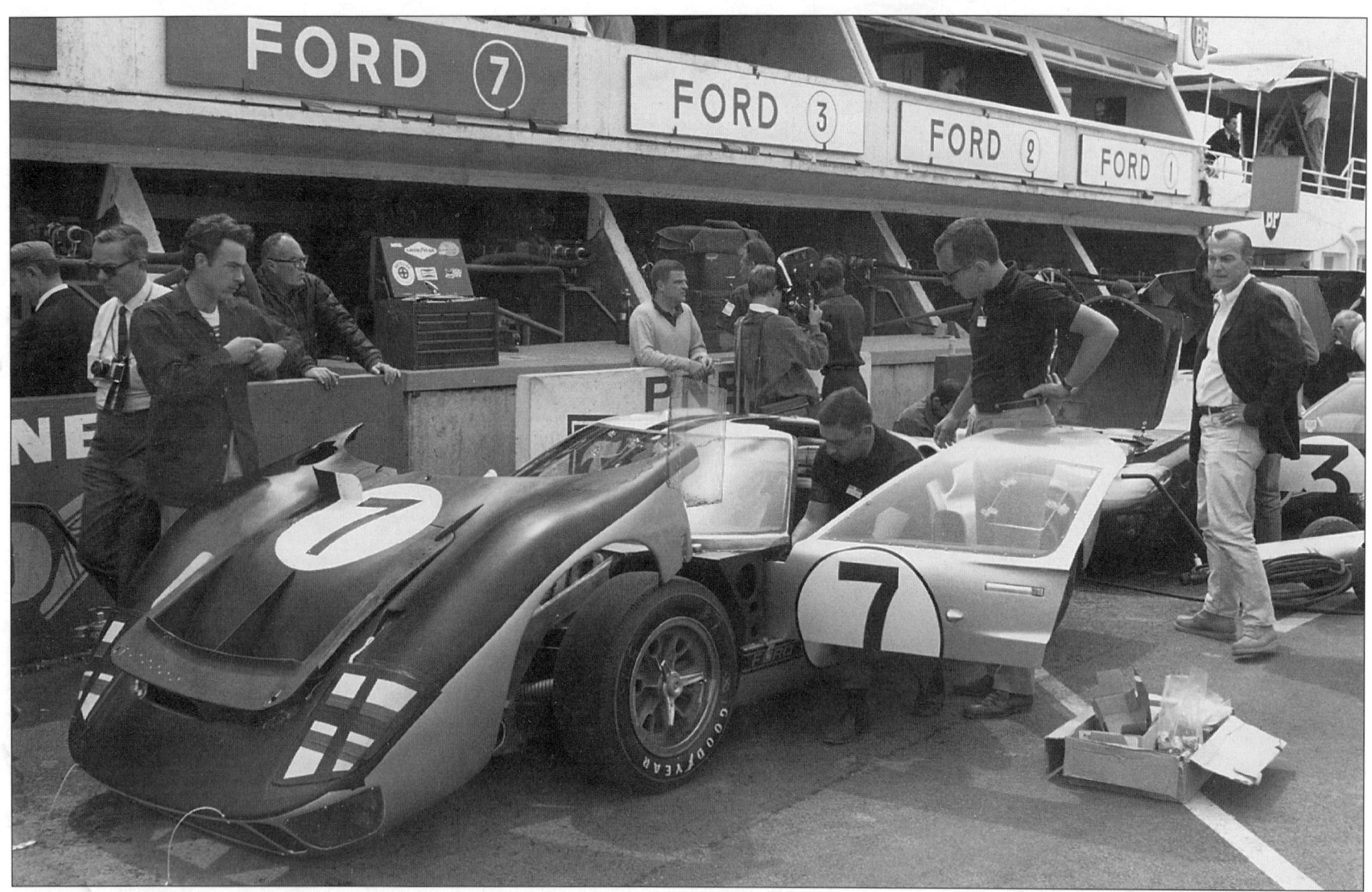

Chassis No. XGT-2, a lightweight-roof car and yet another serial number series. The car was built at Shelby's shop by a crew working for England's Alan Mann. As with other Ford GTs, the front hood tilted forward or could be entirely removed. This photo was shot during Le Mans practice in April 1966. The car ran in the race, with Graham Hill and Brian Muir driving, but did not finish as a front upright failed. Here and during the daylight hours of the race, the headlight covers were taped over to protect them from breakage. On race day, the headlights would be uncovered during the last pit stop before darkness.

Le Mans 1966. Rear deck "snorkels" indicated a MK II-A. It was the Gurney/Grant car. Rather than risk fouling up the plugs with low-speed parking maneuvers, the cars were pushed into position on the grid. The works ran eight GT40 MK II-As. The remaining five spots were taken by independently-sponsored small block MK Is.

Ten minutes before race time. You could barely see the pit structures! Car No. 3, Chassis No. 1047, was seen in pole position behind the three gendarmes. Piloted by Dan Gurney and Jerry Grant, it ran like a locomotive for 17 hours but, in the 18th hour, a head gasket failed.

A solemn moment before the race, when various and sundry VIPs walk down the pit straight with flags. Among the dignitaries was Henry Ford II, who was anything but solemn 24 hours later when Ford GTs finished 1-2-3.

At precisely 4 o'clock the flag dropped. The Ferrari prototypes got off to a slow start.

Two MK II-As chase a Ferrari. The vertical metal boxes seen almost dragging on the pavement were, amazingly, F.I.A.-required storage bins. (These cars were, after all, grand touring cars!) The strategy was to run the MK IIs at a 6200 rpm red line, for the first few hours, and then, to ensure the engines would last the race, progressively step down to lower rpm levels. Dearborn's tests on rolling road dynos indicated engine life of a mere 30 hours, if run constantly at full throttle.

Why, in 1966, was this MK I still running on Borrani wire wheels? Blame it on FAV, it was their entry for Ireland and Rindt. Rindt over-revved it and blew the engine, before Ireland ever got a chance to drive.

The Gurney/Grant MK II-A in the pits. The molded-in rear spoiler that had worked in 1965 was no longer judged sufficient. An additional aluminum spoiler was installed. Note third rear deck scoop for gearbox cooling. Presence of lads in pit lane would indicate this was shot during practice.

The proof of the pudding was in the winning! As the clock struck 4:00 P.M., three Fords approached the finish line ahead of the pack. At the last minute, someone at Ford tried to orchestrate a three-abreast finish. However, word did not get to Bruce McLaren, in the black car, who scooted ahead of the others and robbed Ken Miles of the lead—and a Le Mans victory. Ironically, Ford let the winning MK II, Chassis No. 1046, slip out of their hands. In 1983, the car was found in Belgium and bought by an American collector.

The flag man dropped the flag on McLaren, but missed the chance to drop it on Ken Miles.

Here, he concentrated on the third car about to cross the line, the Bucknum and Hutcherson MK II.

Second place co-driver Denis Hulme waved to the crowd. The Miles/Hulme car looked remarkably unsullied, considering it had just hammered over 2,000 miles at an a average speed of 125.113 mph.

Henry Ford II, who was crowned with Ettore Bugatti's nickname "Le Patron" by virtue of his underwriting of Ford's Le Mans effort, appeared in the victory circle with Bruce McLaren and Chris Amon at either hand. There is no explanation for the presence of London bobbies among the French crowd—it is simply one of the unsolved mysteries of Le Mans.

With the 1966 Le Mans win in hand, Ford decided to rush ahead with the all new car, begun in the fall of 1965, called the J-Car. It was named after Appendix J of the F.I.A. rules. Here the clay model, off of which the fiberglass bodies were molded, took shape.

A Ford modeler sculpted on the side of the J-Car clay model.

The modelers' work extended as far as the J-Car's fixed-in-place seats.

The J-Car chassis was bonded together with glue, then baked in a giant oven by Brunswick Corporation, Muskegon, Michigan. Exhausts were no longer the bundle-of-snakes cross-over-type, as on the small blocks and MK IIs. Instead, tuned headers separated on each side, speedboat-style. A two-speed automatic-type converter transmission was used.

A seldom seen picture of the J-Car engine compartment. When the J-Car was tested at Le Mans in April 1966, Ford was quoting 475 hp at 6200 rpm. Rules required drivers carry a spare tire, for which, owing to the flatback shape of the car's rear body section, there was room.

Here, the first complete J-Car clay model and macho driver. When Ford went to the all-new shape, they disregarded everything they had learned with the MK I and II and inherited a whole new set of problems. "Perry's periscope"—the periscope devised by engineer Homer Perry is not on the car yet. Note Firestone tires, perhaps an inside joke to irk Shelby, who was a Goodyear tire dealer.

Rear view of the first complete J-Car clay model.

Le Mans test weekend, April 2-3, 1966. Probably chassis J1. Someone on the Ford team must have liked unlimited hydroplane racing. How else could you explain the "picklefork" nose?

Note the Thunderbird taillamps. The raised rear spoiler looked like a late addition. "Homer's Folly", the tank periscope, looked useful here. A MK I sits on the grid ahead of it, probably the Ford of France entry. The auto club sticker on the J-Car was someone's idea of a joke. Ford designer Homer C. LaGassey, Jr. reported "de-laminations" occurred after hundreds of laps. In other words, the car came unglued. The chassis was taken back to Los Angeles, where Shelby's crew added metal braces at all angles.

In the pits at Le Mans and on its practice runs (opposite), the J-Car looked formidable. It would appear that the NACA ducts on the front hood had been taped over.

The MK IV was really the J-Car re-bodied. The new body was developed in a rush program, after doubts set in about the aerodynamic stability of the J-Car, following the crash in testing at Riverside that killed Ken Miles. The shape of the MK IV was developed in only a few weeks and mostly arrived at through intuition—Ford officials trying to figure out what caused Miles' crash and then changing it—and, finally, wind tunnel-testing the result.

The MK IV's fastback was more in line with then current race car practice than the "flatback" of the J-Car. What appears to be black string is actually yarn taped on to show how the air flows over the body.

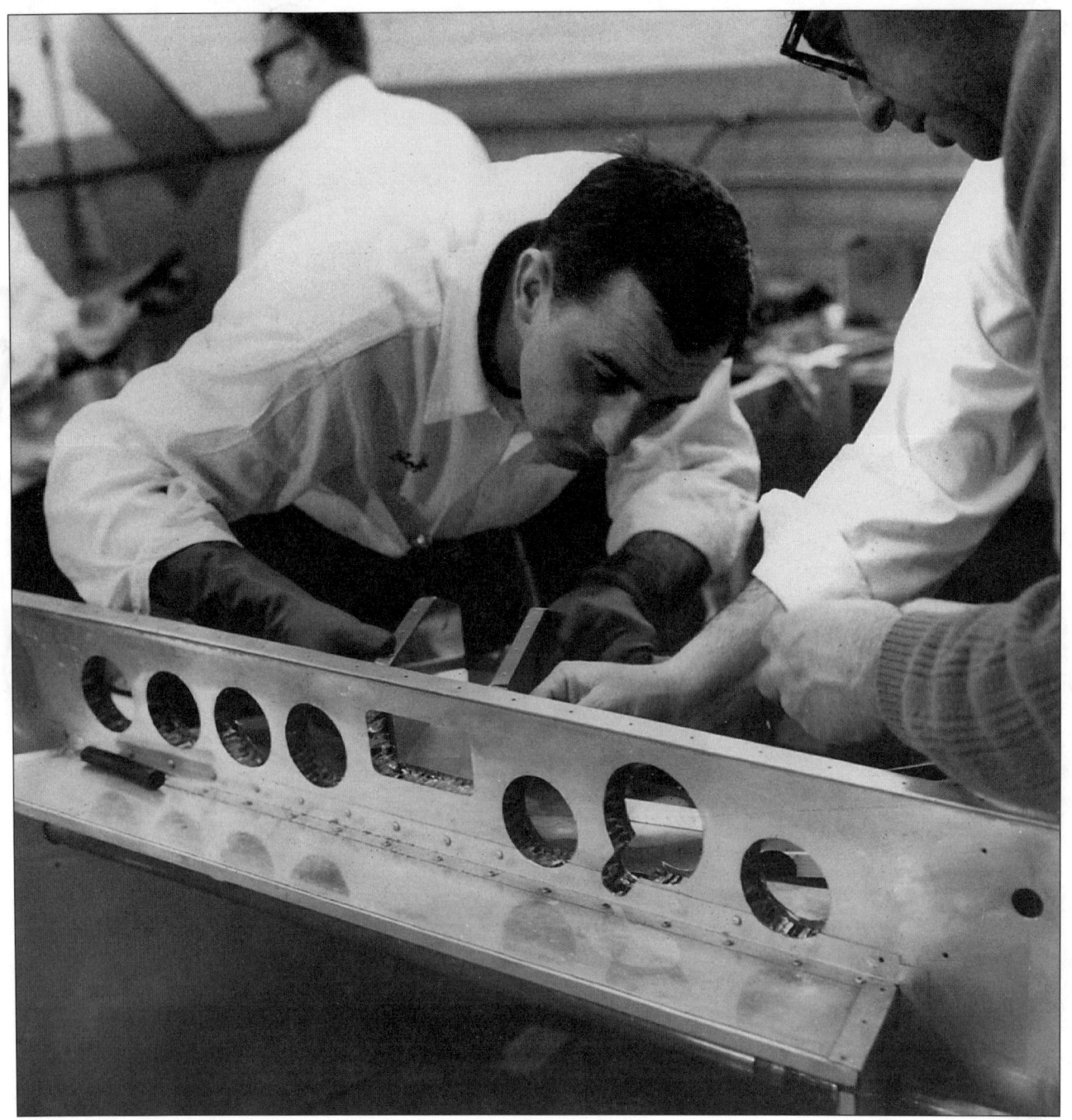

The MK IV chassis was made of aluminum honeycomb sandwich material, derived from aircraft use. Glued together, the tub was baked in an oven to harden the adhesive.

MK IV and J-Car chassis detail: (upper left) the Ford crew shown at work on the tub, before the suspension was fitted; (upper right) Huge horizontal bracket with holes, fixed above the engine, was a brace for "speedboat-style" exhausts; (lower left) a crew member worked on the front suspension mounts; (lower right) a partial view of the completed suspension, brake, and steering systems.

After Miles' accident, Ford went over the MK IV and added additional bracing strips of aluminum, pop-riveted on, to strengthen the chassis. They even added a roll cage. The result? Several hundred pounds more weight, but at Le Mans in 1967 a MK IV crashed and the driver walked away. The vertical tank in the engine compartment was a coolant overflow catch tank. The horizontal rectangular box on the gunwale was the air scoop that led to the oil cooling radiator. Note exhausts were no longer the "boat" exhausts of the J-Car, but cross-overs like those of the MK I and MK II, intended to simulate the effect of a flat crank. The engine was rated at 530 hp. Transaxle was a Kar Kraft T-44 four-speed. The "Gurney Bump", designed to accommodate the driver's height, would indicate that this was the 1967 Le Mans winner.

The MK IV tail was very much "form-follows-function". The fastback was plain, with no air vents on the deck lid as on the MK I, II, III, and Mirage. The open area around the exhausts was similar in size to that of the MK II but more square in shape. Height of the rear spoiler was adjustable.

You have to give Dearborn credit—even though they did not sell many MK I road cars, they tried again with the MK III. Unfortunately, they humped the front fenders higher to accommodate the larger diameter DOT-required sealed beam headlamps, and lengthened the tail to provide more luggage room. Purists say they altered what was already a perfect car. The MK III offered Sebring Mach I mirrors, chrome-framed glass windows, and wire wheels, which Shelby had dropped from competition cars a couple years before. Chrome bumperettes offered laughable protection. They even installed lockable doors, as if anyone would have the nerve to steal a GT40!

The MK III interior. A little more comfort than the first road GT40. Only seven MK IIIs were built, at least two with left-hand drive.

The horizontal luggage box of a MK III. Note aluminum shields that protected the fiberglass deck lid from the heat of exhaust. Note how side glass pops open at back for ventilation.

MK III assembly line, Slough, England. Five of the entire run of seven cars are seen here. Customers included singer Vic Damone and orchestra conductor Herbert von Karajan.

The MK III, at left, and MK I, at right. This side-by-side comparison showed how easily the changing of a few dimensions changed the entire character of the GT40. The MK III had a longer tail for the revised luggage bin.

MK I, left, MK III right. These were restored cars, rolled out for a German magazine's comparison drive in the late 1980s.

Sebring 12-Hours 1967. The MK IV's J-Car legacy can be seen in the cabin shape. The nose and tail were less radical, however. The front air extraction vent was one big vent, moved a bit further aft, with "fences" added, so that wind that flowed over the body would not interfere with air that exited the vent. This bright yellow MK IV, Chassis No. J4, was driven by Mario Andretti and Bruce McLaren, and finished first overall. Foyt and Ruby, drove the only other Ford-sponsored GT40 and finished second in their MK II.

The Andretti/McLaren 1967 Sebring MK IV at speed. Rear view revealed damage to side and rear of the car. Thus bruised and battered, the car still took first at an average speed of 102.293 mph. The MK IV was 300 lb. lighter than the MK II. At the Kingman, Arizona test track, it was clocked at 215 mph on the banked portion of the track.

After the 1966 Le Mans victory, Ford decided to go with a "Made-in-USA" car for the next generation. Thus, John Wyer and his partner, Sir John Willment, founded JW Automotive Engineering. With Len Bailey at the drawing board, they developed the Mirage. It was a narrow-roofed car, based on the GT40 MK I and built to Group 6 rules. This photo was shot at the April 1967 Le Mans trials, where the car was fitted with a 4.7-liter engine. The following month at Spa, it ran its first race. At the 1967 24 Hours of Le Mans, the car ran with a 5.7-liter engine but blew a head gasket and did not finish.

The front air duct on the Mirage was changed, from that of the MK I and MK II-A, to one big duct with no center split. Canard front spoilers looked like an afterthought.

Race day, Le Mans 1967. Homer Perry stood in front of two of Ford's MK IVs. Here and until darkness, the winning car, No. 1, wore headlamp covers that protected the Plexiglas.

Race start. Flag men hastened to clear the course. Fords predominated at the front of the grid, although a bewinged Chapparal 2F somehow inserted itself between Fords No. 2 and 3. Jim Hall, Ford's would-be nemesis, entered two Capparals. Mike Spence, Hall's co-driver, ran second at the two-hour mark when the wing actuator broke, later followed by transmission seal failure. The second 2F dropped out with bum electrics.

MK IV No. 3 on the Ford team looked impressive at the start, when drivers Lucien Bianchi and Mario Andretti took it out. A few hours later, Andretti came in for new brake pads, and returned to the track only to find the brakes grabbed unevenly. He hit the wall and ripped the back end off the car. Another Ford driver, McCluskey, came along in a MK II-B and threw his car into a spin to avoid Andretti. Schessler came next. Result? Three Fords out in one accident.

Seen from the front, there was nothing recognizable from the previous Ford GTs—MK I, II, and III. If anything, the MK IVs borrowed from the shape of the Ferrari P3/4's! This was the winning car at Le Mans 1967. A. J. Foyt and Dan Gurney ran 3,215.5 miles at an average speed of over 135.48 mph. Note the "Gurney Bump" over the driver's compartment.

The winning MK IV trundled along toward the sleepy village of Le Mans at over 200 mph. The shiny aluminum box to the right of the exhaust was the F.I.A.-required "luggage box".

The victory lap is a cherished, although dangerous, tradition, at Le Mans. Dan Gurney and an unidentified official took *the* lap.

Moët and Chandon provided the champagne; Dan Gurney and A. J. Foyt the enthusiam. Le Mans' 1967 victors.

When the F.I.A. came up with new rules that made it impossible to run the Mirage body style, Wyer turned to the rule book. Because over 50 GT40s had been built, the GT40 qualified as a production sports car. Wyer converted two Mirages back to GT40s, and, depending upon the class in which they ran, went to either a five-liter or 5.7-liter engine. Note front canard fins, wideness of body, and higher "duck tail" with add-on rear spoiler.

Dr. Dick Thompson, the Washington DC dentist, lost control of his Mirage near Clearway Corner at Brands Hatch, during the 1967 BOAC International 500 Six Hour Sports Car Race. His only injury was a cut on the cheek! The race was won by a Chaparral, co-driven by Phil Hill and Mike Spence.

Spa 1968. The Wyer campaigning of the GT40s marked the redemption of the original concept, before Ford became obsessed with the MK II big blocks. The Gulf GT40s were small block, and employed the MK I body style with only minor changes. Here you can see additional venting carved into the rear panel. Note grate with oblong holes that replaced earlier louvers on the rear deck.

Another Gulf GT40 at Spa 1968. Note aluminum fender lip, added to the rear, and the slim front canards. This was Chassis No. 1079. In 1968, the Gulf GT40s ran with Gurney-Weslake heads on 5-litre engines that eventually produced 465 hp.

The Gulf GT40s of 1968 and 1969 were not as plain in livery as the Ford Motor Company factory entrants of 1964 through 1967. Paint was light blue with orange trim. Sponsor decals included Firestone, Autolite, Koni, Marchal, and Ferodo. Note the circular hole cut into the toll window, which possibly aided communication with the pit crew. This was the Ickx/Oliver car on their way to victory at Le Mans in 1969.

How strong were GT40s? There were fatalities, that included Walt Hansgen during practice for Le Mans, but several were crashed with only minor injuries to the driver. (See page 117.) This was Chassis No. 1001, a private entry run by C. Lucas Engineering at Silverstone in 1968. The driver lost it in the rain, hit the wall, but survived.

How to get million dollar cars for free? Open a museum. The late Bill Harrah, a dapper casino owner in Reno, Nevada, collected cars by the thousands. When Ford needed storage room and saw some tax write-offs, they donated a MK II-A and a MK IV to Harrah's museum. This was Harrah in the MK II-A, as Homer Perry explained starting procedures. The MK II-A was Chassis No. 1015; the MK IV, No. J-8. The MK II-A was gold with red kock-offs. Note racing door handle. The large circle at the end of the handle made it easier to open the door.

In late 1966, Ford thought they had better squeeze some value out of all the millions they had spent on racing. They drafted Carroll Shelby, Phil Remington, and a few other SAI employees, and set out on the road with grab-bag of cars that included, left to right, a 289 racing Cobra, a GT40, a Cobra Daytona coupe, and R-Model Shelby GT-350. This picture was given to Wallace Wyss by Mr. Shelby, during the Cobra Caravan's stop over in Detroit.

ACKNOWLEDGMENTS

Special thanks to Jack Telnack, Vice President Design, Ford Motor Company, for the photographs of clay models.

Special thanks to Harry Calton for making Ford of Britain photographs available during research for earlier books, and to current employees of Ford of Britain for doing the same again.

Special thanks, also, to Paul Preuss, formerly of Ford Public Relations and Ford's PR man during the battles for Le Mans, for providing artwork through the last 20 years of this editor's research.

Thanks to Homer C. LaGassey, Jr. for the loan of photographs from his collection.

John Clinard and Sandra Badgett of Ford's West Coast PR office are also owed thanks for their support of our research.

Thank you, too, to Bob Nagstad, formerly of Ford's Kar Kraft and later SVO, and former Shelby-American International employees: Phil Remington, John Collins, Jerry Dondio, Phil Ren, and Bob Bondurant for making photographs available from their collections.

Finally, particular thanks to Carroll Smith for reviewing the final draft and assisting in the correction of captions, as well as for the loan of photographs from his collection.

The Iconografix Photo Archive Series includes:

AMERICAN CULTURE

AMERICAN SERVICE STATIONS 1935-1943	ISBN 1-882256-27-1
COCA-COLA: A HISTORY IN PHOTOGRAPHS 1930-1969	ISBN 1-882256-46-8
COCA-COLA: ITS VEHICLES IN PHOTOGRAPHS 1930-1969	ISBN 1-882256-47-6
PHILLIPS 66 1945-1954	ISBN 1-882256-42-5

AUTOMOTIVE

IMPERIAL 1955-1963	ISBN 1-882256-22-0
IMPERIAL 1964-1968	ISBN 1-882256-23-9
LE MANS 1950: THE BRIGGS CUNNINGHAM CAMPAIGN	ISBN 1-882256-21-2
PACKARD MOTOR CARS 1935-1942	ISBN 1-882256-44-1
PACKARD MOTOR CARS 1946-1958	ISBN 1-882256-45-X
SEBRING 12-HOUR RACE 1970	ISBN 1-882256-20-4
STUDEBAKER 1933-1942	ISBN 1-882256-24-7
STUDEBAKER 1946-1958	ISBN 1-882256-25-5
LINCOLN MOTOR CARS 1920-1942	ISBN 1-882256-57-3
LINCOLN MOTOR CARS 1946-1960	ISBN 1-882256-58-1
MG 1945-1964	ISBN 1-882256-52-2
MG 1965-1980	ISBN 1-882256-53-0
GT40	ISBN 1-882256-64-6
FERRARI PININFARINA 1952-1996	ISBN 1-882256-65-4
VANDERBILT CUP 1936 & 1937	ISBN 1-882256-66-2

TRACTORS AND CONSTRUCTION EQUIPMENT

CASE TRACTORS 1912-1959	ISBN 1-882256-32-8
CATERPILLAR MILITARY TRACTORS VOLUME 1	ISBN 1-882256-16-6
CATERPILLAR MILITARY TRACTORS VOLUME 2	ISBN 1-882256-17-4
CATERPILLAR SIXTY	ISBN 1-882256-05-0
CATERPILLAR THIRTY	ISBN 1-882256-04-2
CLETRAC AND OLIVER CRAWLERS	ISBN 1-882256-43-3
ERIE SHOVEL	ISBN 1-882256-69-7
FARMALL F- SERIES	ISBN 1-882256-02-6
FARMALL MODEL H	ISBN 1-882256-03-4
FARMALL MODEL M	ISBN 1-882256-15-8
FARMALL REGULAR	ISBN 1-882256-14-X
FARMALL SUPER SERIES	ISBN 1-882256-49-2
FORDSON 1917-1928	ISBN 1-882256-33-6
HART-PARR	ISBN 1-882256-08-5
HOLT TRACTORS	ISBN 1-882256-10-7
INTERNATIONAL TRACTRACTOR	ISBN 1-882256-48-4
JOHN DEERE MODEL A	ISBN 1-882256-12-3
JOHN DEERE MODEL B	ISBN 1-882256-01-8
JOHN DEERE MODEL D	ISBN 1-882256-00-X
JOHN DEERE 30 SERIES	ISBN 1-882256-13-1
MINNEAPOLIS-MOLINE U-SERIES	ISBN 1-882256-07-7
OLIVER TRACTORS	ISBN 1-882256-09-3
RUSSELL GRADERS	ISBN 1-882256-11-5
TWIN CITY TRACTOR	ISBN 1-882256-06-9

RAILWAYS

CHICAGO, ST. PAUL, MINNEAPOLIS & OMAHA RAILROAD	ISBN 1-882256-67-0
GREAT NORTHERN RAILWAY 1945-1970	ISBN 1-882256-56-5
MILWAUKEE ROAD 1850-1960	ISBN 1-882256-61-1
SOO LINE 1975-1992	ISBN 1-882256-68-9

TRUCKS

BEVERAGE TRUCKS 1910-1975	ISBN 1-882256-60-3
BROCKWAY TRUCKS 1948-1961*	ISBN 1-882256-55-7
DODGE TRUCKS 1929-1947	ISBN 1-882256-36-0
DODGE TRUCKS 1948-1960	ISBN 1-882256-37-9
LOGGING TRUCKS 1915-1970	ISBN 1-882256-59-X
MACK® MODEL AB*	ISBN 1-882256-18-2
MACK AP SUPER DUTY TRUCKS 1926-1938*	ISBN 1-882256-54-9
MACK MODEL B 1953-1966 VOLUME 1*	ISBN 1-882256-19-0
MACK MODEL B 1953-1966 VOLUME 2*	ISBN 1-882256-34-4
MACK EB-EC-ED-EE-EF-EG-DE 1936-1951*	ISBN 1-882256-29-8
MACK EH-EJ-EM-EQ-ER-ES 1936-1950*	ISBN 1-882256-39-5
MACK FC-FCSW-NW 1936-1947*	ISBN 1-882256-28-X
MACK FG-FH-FJ-FK-FN-FP-FT-FW 1937-1950*	ISBN 1-882256-35-2
MACK LF-LH-LJ-LM-LT 1940-1956 *	ISBN 1-882256-38-7
MACK MODEL B FIRE TRUCKS 1954-1966*	ISBN 1-882256-62-X
MACK MODEL CF FIRE TRUCKS 1967-1981*	ISBN 1-882256-63-8
STUDEBAKER TRUCKS 1927-1940	ISBN 1-882256-40-9
STUDEBAKER TRUCKS 1941-1964	ISBN 1-882256-41-7

* This product is sold under license from Mack Trucks, Inc. All rights reserved.

The Iconografix Photo Archive Series is available from direct mail specialty book dealers and bookstores worldwide, or can be ordered from the publisher. For book trade and distribution information or to add your name to our mailing list contact:

Iconografix
PO Box 609/BK
Osceola, Wisconsin 54020 USA

Telephone: (715) 294-2792
(800) 289-3504 (USA)
Fax: (715) 294-3414

MORE GREAT BOOKS FROM ICONOGRAFIX

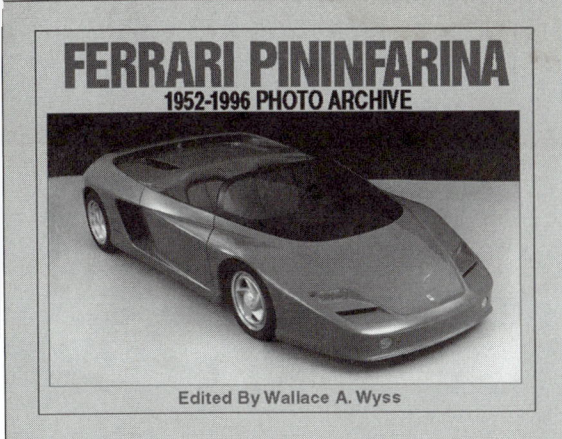

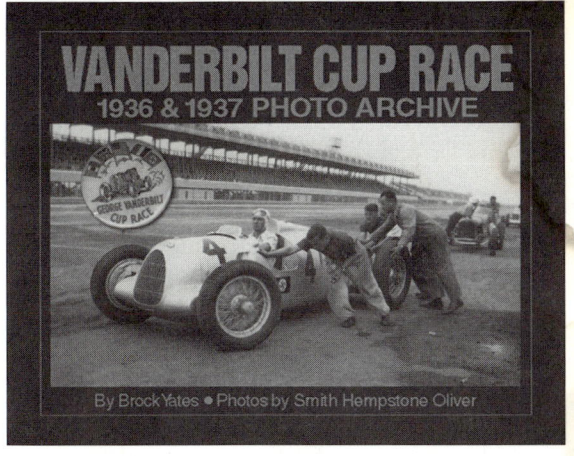

FERRARI PININFARINA 1952-1996
Photo Archive ISBN 1-882256-65-4

VANDERBILT CUP RACE 1936 & 1937
Photo Archive ISBN 1-882256-66-2

LE MANS 1950 PHOTO ARCHIVE THE BRIGGS CUNNINGHAM CAMPAIGN
ISBN 1-882256-21-2

SEBRING 12-HOUR RACE 1970
Photo Archive ISBN 1-882256-20-4

AMERICAN SERVICE STATIONS 1935-1943 Photo Archive
ISBN 1-882256-27-1

MG 1945-1964 Photo Archive
ISBN 1-882256-52-2

MG 1965-1980 Photo Arhive
ISBN 1-882256-53-0